# Mindful thoughts for
# **STARGAZERS**

First published in the UK in 2019 by
*Leaping Hare Press*

An imprint of The Quarto Group
The Old Brewery, 6 Blundell Street
London N7 9BH, United Kingdom
**T** (0)20 7700 6700  **F** (0)20 7700 8066
www.QuartoKnows.com

© 2019 Quarto Publishing plc

All rights reserved. No part of this book may be reproduced or transmitted in any form or by any means, electronic or mechanical, including photocopying, recording, or by any information storage-and-retrieval system, without written permission from the copyright holder.

British Library Cataloguing-in-Publication Data
A catalogue record for this book is available from the British Library

ISBN: 978-1-78240-766-9

This book was conceived, designed and produced by
*Leaping Hare Press*
58 West Street, Brighton BN1 2RA, UK

Publisher: *Susan Kelly*
Editorial Director: *Tom Kitch*
Art Director: *James Lawrence*
Commissioning Editor: *Monica Perdoni*
Editor: *Claire Saunders*
Illustrator: *Lehel Kovacs*
Editorial Assistant: *Niamh Jones*

Printed in China

1 3 5 7 9 10 8 6 4 2

# Mindful thoughts for
# STARGAZERS
*Find your inner universe*

**Mark Westmoquette**

*Leaping Hare Press*

# Contents

INTRODUCTION **Journey to the Inner Universe** — 6

The Mystery of What's Up There — 10

Cloudy Nights — 16

Twinkling Stars — 22

Finding Meaning — 28

Navigating — 34

Red Sky at Night, Shepherds' Delight — 40

Rhythms and Cycles — 46

The Moon — 52

Under Mother Nature's Protection — 58

The Planets — 64

Artificial Stars — 70

Polluted Skies — 76

Through the Looking Glass — 82

| | |
|---|---|
| Time and Space | 88 |
| A Sprinkling of Stardust | 94 |
| A Photon's Journey | 100 |
| Is Space Really Empty? | 106 |
| Shooting Stars | 112 |
| Entering Silence | 118 |
| Zoom Through the Universe | 124 |
| The Most Exquisite Jewels | 130 |
| How Many Stars Are There in the Sky? | 136 |
| Is There Anybody Out There? | 142 |
| The Overview Effect | 148 |
| Bringing It Back to Earth | 154 |
| ACKNOWLEDGEMENTS | 160 |

INTRODUCTION

# Journey to the Inner Universe

---

At 18, I decided to go to university to study astrophysics. Looking back, I can see my decision to study astronomy was partly motivated by wanting to escape, after a very traumatic childhood, to the furthest reaches of the Universe. After my undergraduate degree I was swept along into a PhD and ended up spending ten years doing research with my head buried in space.

During that time, I took up yoga and later Zen meditation. I remember my intellectual mind initially being quite resistant to all the what seemed like woo-woo nonsense in these Eastern practices, but my subconscious brought me back to the classes week after week. Over time, I rediscovered the forgotten lands of my body and the subtle contours of feeling and sensation. With mindfulness, I started to see the patterns of behaviour and ways of being I had adopted as methods for avoiding any more pain. With guidance from my Zen teacher and my psychotherapist I began to understand how these very patterns had now become the source of my suffering.

Meditation showed me how to allow strong emotions to arise, without feeling like I had to do anything about them or find out why they were there. Slowly I realized I didn't need to escape into space any more. I experienced first-hand the Buddha's teaching on suffering. Suffering arises, he said, because when we experience a difficulty we either wish it wasn't there, or we just deny or ignore it. I was very happy to find this has been codified into an equation: Pain × Resistance = Suffering.

A few years ago, I decided to end my astronomy research career and take up teaching yoga and meditation full-time. It has taken me quite some time to see how astronomy – that often-cold realm of maths and physics that I had escaped into – can actually be a tool for connecting us more deeply with life here on Earth. And that's the intention of this book: to explore various aspects of the stars, the planets and the Universe to help us live with more presence and awareness of our own two feet on the ground. The first half of the book focuses on the practical side of mindfulness – dealing with the mind and thoughts, navigating our life and dealing with emotions. The second half looks at the bigger picture, including the deeply interconnected nature of the Universe, life and death.

# The Mystery of
# What's Up There

I used to think when I was working in astronomy that if we stopped and actually thought about what we were studying we'd fry our circuits on a daily basis. Just think of the distances, sizes and timescales involved. The nearest star after the Sun, Alpha Centauri, is 42 trillion kilometres away – a totally inconceivable distance. It takes light almost four and a half years to travel that far. This is why astronomers call this distance 4.4 light-years, four being a much easier number to deal with than 42 trillion. Astronomers come up with all sorts of units for distance, time and energy, just to make them more comprehensible and less mind-bending to work with.

The night sky is as mysterious and awe-inspiring to us modern humans as it has been to any animal that has ever had the capability of looking up and wondering. When we gaze up at the stars we are sharing that view with all our ancestors. Imagine an early human out on the savannah, stepping away from the evening fire after a meal of antelope and foraged vegetables, glancing up at the perfectly dark sky and being totally captivated. Black that's blacker than black, spaciousness extending to infinity, and tiny, bright sparks dotted everywhere, twinkling silently. Eyes wide, head tilted, and just the sound of his or her own quiet breathing and the crackling of the fire. And maybe later, once everyone else has gone to sleep, this man or woman's mind starts questioning: What's up there? What are all those bright dots? How far away are they? Is that where we come from? What kind of power (or deity) keeps it all going?

## A MEDITATION ON LOOKING

These days it is difficult to find a place quite as dark as that which the prehistoric human experienced. This is a shame, because when it is dark right down to the horizon you cannot help but be completely drawn into the vista. Living today, we see those same tiny, bright sparks twinkling silently, piercing the black. Gazing skywards can be a meditation itself – a meditation on looking. And when you do look, what happens in your body? How do you feel it? For me, as my attention is drawn upwards, I feel a sense of wide expansiveness, spaciousness and calm. My peripheral vision slowly activates and my eyes feast on the increasing number of faint stars which reveal themselves to me. The familiar constellations are overwhelmed by the sheer number of fainter stars obscuring their patterns. I feel totally immersed. Until I realize my neck is hurting and I'm freezing cold . . .

There is no other view in our human experience that we can share with so many others back and forward through time. Earthly landscapes have changed over the millennia (eroded, slid under the sea or been built on), but the night sky has remained (more or less) the same. The oldest star chart we know of is over 32,000 years old, etched into a sliver of mammoth tusk found in 1979 in a cave in Germany.

It shows a constellation that looks remarkably like Orion. If humans make it to the year 100,000, the Ursa Major constellation (the Big Bear, or Big Dipper) will have only changed slightly, looking more like a large kitchen knife than a saucepan.

## ASKING THE BIG QUESTIONS

The majesty and mystery of the stars above which stirred that Stone Age human to spend precious time and energy carving Orion into a mammoth tusk, also, over time, inspired humans to build the astrologically-aligned great pyramids of Egypt and of Mesoamerica, monuments such as Stonehenge in England and Brú na Bóinne in Ireland, and cities like Machu Picchu in Peru and Teotihuacan in Mexico. And these days our thirst to understand more of this mysterious Universe motivates the funding of the world's greatest astronomical observatories such as the VLT (Very Large Telescope) and ALMA (Atacama Large Millimeter Array) in Chile and W. M. Keck in Hawaii, as well as the Hubble Space Telescope. We still ask many of the same questions: What's up there? How far away are they? What energy keeps it all going? Is that where we came from?

Enshrining astronomical alignments in architecture extends into the modern era too. The streets of Manhattan in New York, known as Manhattanhenge to some, are aligned such that at the winter and summer solstice, the Sun can be viewed straight down many streets. Similarly, the main shopping centre in Milton Keynes in the UK is designed so that the central boulevard is aligned with the Sun at summer solstice sunrise. The Shanghai Financial Centre skyscraper was built with an opening that aligns with the major lunar standstill point (the point where the moon stops getting any higher in the sky – analogous to the solar solstice). But however much we look for the answers to what's out there and why, and however inspired we are to express our inquisitiveness in architecture, ultimately, true satisfaction and happiness is only found when we look deeply within.

# Cloudy
## Nights

---

For the avid stargazer, the weather can be very frustrating. If it's not partially overcast it's fully overcast, or there's a subtle high cirrus that dulls the pin-points of light we're interested in. That is why the best observatories are situated on high mountains (or in space) – to get above the clouds.

I will never forget being up on Mauna Kea in Hawaii after a night of working at the telescopes. As dawn broke I looked down the mountain to see a sea of cloud blanketing the island below. We had been happily above it all night.

I have had my fair share of frustration with the weather. On one occasion I travelled halfway round the world to work on a big telescope, but instead of spending the night in the control room, guiding the telescope to my list of targets, I was stuck in the lounge playing table tennis with my colleagues. With heavy eyelids we waited for the cloud to clear and the wind to die down.

Frustration is an interesting emotion. It is a kind of bubbling, heated restlessness, often accompanied by thoughts like, 'Why is it like this? Why me? Why now?' Left unattended it can suddenly erupt into anger or rage, or manifest itself in unskilful, irritable comments. Buddhist teachers call this state 'dis-ease'. But to be angry at the weather for being cloudy is like being angry at the grass for growing. It is just doing what it does.

We get angry because things don't go 'our way'. I might think to myself, 'I've spent hours writing the proposal, planning my observations, flying across the continent and staying up all night for this?? Clouds?!' Or as a back-garden stargazer I might say, 'I've been watching the forecast all week and this was supposed to be the clearest night. When I set my alarm for 3am so I could go out to catch the sky at its darkest I expected it to be totally clear – but it's not!'

## THE THREE POISONS

The forces of want (for example, wanting the clouds to clear), of aversion (hating clouds or the cold) and of delusion (running 10 minutes on the treadmill waiting for the clouds to clear, and so feeling we deserve a chocolate bar) are what the Buddha called the 'three poisons'. They are our source of suffering. We get stuck when we start wanting or wishing things to be different from what they actually are, right here and now. The further our wish is away from reality, the more we suffer. Sure, we can move to change things in the future, but right now, this is how it is.

Sadly, people waste so much energy wanting things to be different. Sometimes, it can be difficult to even notice that we are doing it – it can be quite subtle. This is where mindfulness comes in. First, we might notice or realize what we were feeling a day later; then, with practice, we might realize it an hour later, then a minute later, until we're able to see ourselves wanting at the time of wanting. The Buddha used the image of a stuck wheel to describe this state of dis-ease or dissatisfaction. He taught that when we learn how to see and accept things as they are, then our wheel frees up, life frees up. And all that energy that we were using wanting and wishing becomes free, so that we can use it for something fun.

## APPRECIATING THE LITTLE THINGS

Let us go back to our cloudy night in the back garden. We have been out for half an hour since the alarm went off, trying to identify some of the lesser-known constellations through the shifting cloud, and we're frustrated. We notice our frustration and allow it; we try not to chastise ourselves for feeling like this. Life is the way it is, and no amount of wishing can change that. As soon as we let go of wanting things to be the way we want them to be, there is room for noticing other things. Perhaps the way our breath condenses, forming swirling patterns in the still night air; perhaps a prowling cat or badger on their nightly hunt; perhaps just the stillness and quiet of this time of the morning. When we let go of our rigid expectations and needs, it allows us to be open to the unexpected, and often very beautiful, little things in life. If the sky had been clear, we would have been engrossed in our star charts, and might have missed the sparkling bright eyes of the fox gazing at us from the gap in the hedge.

# Twinkling Stars

One of David Hockney's most famous paintings is of a man standing by an outdoor pool in Los Angeles. He has painted the blue wavy water in the pool in such a way that it actually looks like it is moving – just as it does in real life. Objects underwater appear to shimmy around because of refraction. Even if what is at the bottom of the pool is totally still, it appears in a constant, shimmering motion as the surface ripples and moves. This is exactly why stars twinkle. Out in space, nothing twinkles, but as soon as starlight enters the atmosphere it begins to be bent (refracted) ever so slightly this way and that by the different pockets of air.

You see the same effect in a mirage over the surface of a hot road, and you feel those pockets of air as turbulence when you fly.

In astronomy, twinkling is called scintillation, and the stillness of the atmosphere is called 'seeing'. Where I live in London, on a clear night I can clearly see the restless twinkling of the five stars that are visible beyond the light pollution. On a warm summer's night, they get even more restless as the air rises off the hot tarmac, causing the air to slosh around just like the water in a swimming pool. This is another reason why telescopes are built on high mountains. The atmosphere here is thinner, meaning there is less chance of cloud and the 'seeing' is better. When astronomers go on an observing run on a big telescope, they all have their fingers crossed for good seeing. Of course, cloud and seeing doesn't matter for radio telescopes. One of the UK's most productive radio telescopes at Jodrell Bank is at sea level just outside Manchester – one of the rainiest spots in the country.

## CLOUDY MIND, CLEAR MIND

When we meditate, our concentration can also suffer from bad 'seeing'. You might say our focus 'twinkles' as we are bombarded by endless thoughts, impressions, perceptions, memories and ideas. How do we deal with these distractions? The first thing is to realize that everyone gets them. It is not just you. When we stop and take a moment to look inside our minds, it can feel like complete chaos. That's fine. The brain produces thoughts, just like the intestines digest our food. That is its job. Mindfulness is not about getting rid of these thoughts; instead, what it teaches us is how to change our attitude or relationship to them.

If we envisage a thought like a cloud in the sky, then we know there are times when there are lots of clouds and times when there are few. Sometimes it is totally overcast – our mind is so busy that thoughts are crowding in one after another with no space in between. Maybe it is like that for you your whole working day. Sometimes there are just a few white fluffy clouds scudding overhead. Sometimes the sky clears completely. But having a clear mind is as rare as a blue sky in a rainforest; it is lovely when it happens, but chasing it gets us nowhere. Our job is to accept and allow things just as they are. Let the mind be as cloudy or clear as it is right now, and bring an attitude of kindness and patience to your experience. Even on the cloudiest of nights, the stars are still shining. Even when our mind is racing, we can allow that busyness and acknowledge that it is fine.

People liken mindfulness to a trip to the gym for your mind. Every time we get involuntarily whisked away from our focus and start following a train of thought, we gently guide our awareness back to the here and now. There is no need to try and work out why you were distracted or to chastise yourself. Just come back to now.

## MEDITATIONS ON A STAR

Here is something you can try the next time you are looking up at the night sky. Pick a star and focus all your attention on it. You might want to look slightly to the side of the star, as the very centre of your retina is less sensitive in low light. Let the experience of looking at the star be so intent that there is no room for anything else. Let 'you' looking at the 'star' become 'you-star'. This is equivalent to a narrowing, focusing type of meditation of the kind that allows the mind to quieten down into a state of peace and tranquillity. Now try softening and broadening your gaze, and become aware of your peripheral vision. Let as many other stars, clouds, trees and buildings come into your awareness as you can. This is akin to a broad, open kind of meditation that helps us develop kindness and patience with our experience.

# Finding
# Meaning

---

As humans, we crave meaning in what we experience around us. One of the first words that children learn, and then repeat incessantly, is 'why?' As a species we have always wanted to understand what is happening around us, perhaps so that we can better judge the effect it might have on our lives. For the vast majority of our history, when humans looked up to the night sky we were faced with such deep mystery that it was only with stories and fables that we could make any sense of it at all.

The source of many of the constellation names used in modern European languages are the myths of the ancient Greeks. Many of these stories contain moral and ethical teachings to help us remember to live with virtue and compassion. There are ancient stories, too, from other cultures.

## THE STORY OF TWO STAR-CROSSED LOVERS

One beautiful myth from Japan tells of the Milky Way and how it separates the two brightest stars in the sky. The weaver princess Orihime was the daughter of the Heavenly Emperor Tentei and wove beautiful clothes by the bank of the Tennogawa (literally 'heavenly river', what we call the Milky Way). In a meeting arranged by Tentei, Orihime met Hikoboshi, a cowherd boy who worked on the other bank of the river. They immediately fell in love, got married and forgot their work.

Over time the god's clothes ended up in tatters and cows strayed all over heaven. In anger, Tentei separated the two lovers across the Tennogawa and turned them into Altair and Vega, two of the brightest stars visible from the northern hemisphere. Tentei allowed the two to meet only on the seventh day of the seventh month, when a flock of magpies would fly up to create a bridge across the Milky Way. This event is celebrated in Japan today on 7 September in the festival called Tanabata. In the celebrations, people write their innermost wishes onto a slip of paper, tie it onto a tree branch or twig and place it into a river to be carried off.

## YOUR LIFE'S OWN STORYLINES

What about our own stories? People say one of the functions of the brain is to create meaning out of our experiences to help us learn. Out of those meanings come storylines. Some of these are useful, but in my experience most are not. They serve to limit and fix us into habits and patterns of thinking. For example, you are told as a seven-year-old that you are not good at art and all your drawings of the Sun, moon and stars look like paint chaotically flicked onto the page. Because of its emotive charge, this comment turns into a storyline that is played internally again and again through your later childhood until it becomes fixed into a belief: 'I'm bad at art'.

Mindfulness helps us to see things more clearly and objectively. By applying attention to our feelings, emotions and thoughts again and again, with acceptance and without judgement, we start to see these storylines for what they are: just stories. Some traditions use the question 'who am I?' as a focus for meditation and a way of punching through these storylines. Our psyche is like a huge onion, and using that question we start to peel off the layers one by one. Each answer we come up with may be partly true, but it is not the whole truth – there's more. What would happen if we let go of these storylines that serve to dam the river of our life, and instead allow this beautiful river to run through us into every action and interaction?

To me the meaning of the Tanabata myth is one of inner alchemy or transformation that represents the path of self-enquiry. The black and white of the magpies are the yin and yang, and the boy and girl are our inner masculine and feminine sides. Most of us live in this world of the opposites: I am a man/woman, happy/sad, tense/relaxed, rich/poor, and so on. But there is another way of being that doesn't involve these opposites. The meaning of the word 'yoga' is to unite. Yoga is the practice of bringing together our opposites into a place of oneness. There's no longer me here and the big, bad Universe out there; there's just 'me-Universe'. In Tanabata, the meeting of the lovers and fulfilling of the wish represent the coming together of our being into the oneness of the river of life. In that place where there are no opposites, who are you?

# Navigating

Traditional navigation before the age of compasses and sextants required a deep knowledge of the night sky and how it changed with location, time of night and season. Records of people navigating using the stars date back to Greece in the eighth century BCE, but humans and animals have been using the night skies to find their way since the dawn of time.

Fish, animals and insects have evolved to use the Sun and stars to navigate. Some animals, such as bats and sea turtles, can even sense and use the Earth's magnetic field to orient themselves. In the thirteenth century, the Māori people migrated to New Zealand from eastern Polynesia across vast distances of open ocean without any modern navigational technology such as a sextant, compass or clock. Around the same time, a group of Tahitians made the incredible 4,000-kilometre journey to settle the Hawaiian Islands. In 1976, this voyage was successfully recreated using traditional Polynesian navigation techniques, again with no modern navigational aids. These Pacific migrations were the last step in humankind's colonization of the whole planet, and could only have been done by people with courage and a profound understanding of the stars.

## FINDING OUR WAY

In the northern hemisphere, navigation starts with identifying north through finding Polaris, the North Star. It is the brightest star in Ursa Minor (the Little Bear, or Little Dipper), and it's only a happy coincidence that it (currently) aligns with north so well. In the southern hemisphere it is more complicated. Navigators should first look for the bright constellation Crux, the Southern Cross, then extend a line through the long axis of the cross to about four-and-a-half times its length to find south.

Establishing our direction is one thing, but what about our position? Again, this is easier in the northern hemisphere: the height of the North Star above the horizon equals our latitude. To work out our longitude we also need to know the local time (which in the days before clocks was not that easy). The British only managed to establish their successful ocean trade and international warmongering in the nineteenth century once they had invented a chronometer that could work accurately on a rolling ship. Every time they sailed back to port in London they reset their timepieces to the famous clock at Greenwich Observatory.

## GOING WITH THE FLOW

Given the intimate knowledge needed and the lack of accurate instruments, navigation was as much an art and a spiritual practice as it was a science. This situation hasn't changed when it comes to navigating our inner world. There are no instruments or calculations to help us find our way through the often wild waters of our emotions, relationships, loves and losses. It can feel like we are caught in a river, being carried downstream whether we like it or not. Many people spend a great deal of effort trying to resist this flow of life, attempting to swim against the current or back upstream. In these times it can feel like everything is a struggle and everyone is against you. The fact is, we have far less control over the course of our lives than we think. Maybe we can tweak our route here and there, so that we end up down one tributary rather than another, but we will continue to travel downstream. However, if we can let ourselves go with the flow, rather than resist it, everything becomes easier. The problem comes when we want things to go a certain way. Our wants and wishes are the paddles we use to try and go against the current.

## NAVIGATING THE 'MIDDLE WAY'

The Buddha's advice on navigating life was to follow the 'middle way'. This means finding the middle ground between the extremes – between being too lazy and too forceful, too heavy and too light, too serious and too relaxed. But to find this middle ground we first need to know the edges. Many of our teenage years are spent finding and learning these, but the process is something we refine throughout our life. We need direct, first-hand experience of these extremes to really know them. Then they can act as our navigational stars.

With mindfulness, we can sense when we are heading away from the centre and towards one extreme or the other. Maybe we will notice this very grossly at first, but with time and practice we can recognize the more subtle course changes. Navigating this middle way can feel a little like trying to walk on a knife-edge. My Zen teacher's own teacher said that looking back over his life of practice, he felt like a drunk walking down a path. One moment he would be stumbling into the hedge, and as soon as he had picked himself up he would be off onto the grass the other side. The point is, each time we pick ourselves up again and try once more to find that middle way.

# Red Sky at Night, Shepherds' Delight

---

Looking skywards in the evening at a beautiful pink sunset, we might recall the common expression echoed in many cultures: 'red sky at night, shepherds' delight'. I was surprised to find there is some technical truth to this old adage. When the white light of the Sun passes through the atmosphere it hits dust particles and water droplets. Blue light is scattered more easily than red, leaving the Sun itself to appear depleted of blue (and thus a yellowy colour) and the sky to look blue everywhere. At sunset, the Sun's rays glance through a thick column of atmosphere, and the blue gets scattered even more, changing the colour of the sky from yellow to pinky-red. A high concentration of dust particles in the atmosphere usually indicates high pressure and stable air – hence the shepherds' delight.

Weather also affects space. As I write this, today's solar wind speed is 543 kilometres per second with a density of 6.7 protons per cubic centimetre. For the sixth day in a row, the Sun has no sunspots, indicating that at this moment we are heading into the quiet phase of the solar cycle, called Solar Minimum. I find it amazing that there are websites that give you so much information about the weather in space. Why would we care? Increasingly we are becoming a technology-dependent culture. Phenomena such as solar flares (huge explosions on the Sun's surface) can interfere with or damage satellites or cause dangerous surges in long-distance electricity cables. The Earth's magnetic field protects us on the surface from the worst of this, but if we ever want to travel long distances in space then knowing the weather in space will be of the upmost importance.

Along with the protons of the solar wind, the Sun also emits huge numbers of neutrinos. Neutrinos are tiny, inert subatomic particles that were first postulated in 1930 and only discovered in 1956. The vast majority pass straight through Earth without interacting with it. In fact, there are 50 billion neutrinos passing through our bodies every second. Which means about 300 billion will have gone through you by the time you finish reading this sentence – all without harm. Not all neutrinos come from the Sun. Just occasionally there might be a neutrino produced in a distant supernova whizzing through your left arm.

## READING OUR INTERNAL WEATHER

On a retreat some years ago, we started each day with a group 'weather check', where we reported on our internal weather at that moment. There were responses like, 'This morning I feel a bit misty', or 'I'm totally overcast'. Practising mindfulness offers us the opportunity to develop a finely tuned ability to read the internal weather of our being. What's your internal weather like right now? As I sit here writing this, mine is like a cool, bright summer morning with a sense of excitement about the day ahead. But I know it won't always be like this. Mindfulness of our internal weather shows us that there is an ever-shifting 'feeling-tone' in our body which is only sometimes bubbling and excited, and other times dull, foggy, dark, overcast, clear, washed-out or brooding.

## THE FOUR FOUNDATIONS

When the Buddha taught mindfulness, he spoke of what he called 'the four foundations', or areas of awareness: the body, our sensations, the mind and mind objects. The body means things like posture, movement, the physical space of the body or the processes of the body. Sensations are the things going on in the body, such as tingling, temperature or pressure. The mind includes consciousness and the constructs of the mind. And mind objects are the things that go on in the mind, like thoughts, memories, ideas and perceptions.

It is perfectly possible to practise mindfulness of the mind and mind objects, but the danger with this is that it is very easy to be carried away into thoughts and worries about the past or future. Being mindful of the body and its sensations is simpler. If we notice a tingling in the left palm, then we just notice the tingling. If a storyline about the tingling does arise, we can let it go and come back to the tingling. If it gets stronger, that's fine; if it goes away, that's fine too.

The Buddha taught that mindfulness in any one of these four domains has the potential to take us all the way to enlightenment. So, let us make it easy for ourselves and focus on the physical body and sensations. In this way we can develop an extraordinary sensitivity to our shifting internal weather, whether that be on the national (whole body) or local (individual body part) scale.

# Rhythms and Cycles

There is something very comforting about the regularity and predictability of the night sky. You can be sure that as people go to bed on a winter's night in the northern hemisphere, Orion and Gemini will be wheeling up into view. Or if you are in the southern hemisphere, Lyra and Sagittarius will be just rising. So many of us lead such busy, stress-filled lives, that winding down in time for bed can be quite a challenge. For those who lie awake sleepless into the small hours, these familiar patterns can be a source of reassuring, nocturnal companionship. If you find yourself lying awake with to-do lists and ideas whizzing endlessly around your mind, try opening your bedroom curtains and letting your gaze draw outwards towards the dark sky.

Notice the rhythm of your breath, the rhythm of your heartbeat, the sway of the trees and the steady twinkling of the stars.

Hectic lives mean that our daily rhythms and routines can easily get forgotten. For many, maintaining a steady daily routine needs effort and attention. Getting up at the same time, going to bed at the same time and cooking at a reasonable hour keeps us on a steady keel. If only for that reason, having a friend to stay for a few days or enduring building works at home are challenging. I met a nurse recently who had just worked a nightshift followed by a dayshift three times in a row. I can't imagine how her body clock copes with that.

## OBSERVING THE RHYTHM OF THE SKIES

Day by day, the rising time of each star shifts ever so slightly, as does the time of the sunrise and sunset. That shift is very precise – so much so that we have to add a leap second to our clocks every few years to bring them in line with the heavens. The Sun and moon follow their own particular rhythms, and rarer events such as meteor showers and eclipses add a certain syncopation to the heavenly beat.

Observation of the sky was first undertaken as a way of recording its rhythms and cycles, in order to create a calendar. For most of human history, it has been our only measure of time. In ancient times, astronomers were priests, and astronomical calculations were often preciously guarded secrets. In ancient China, the main responsibility of political power was to keep the Earth in harmony with the sky. This so-called 'Mandate of Heaven' meant that astronomers had influence over daily life as well as major political strategies. In ancient Greece, a solar eclipse around the year 600 BCE was predicted by the philosopher Thales and was interpreted as an omen. It interrupted a battle between the Medes and the Lydians and brought it to a truce.

In many cultures, being able to predict astronomical events meant power. The *Atharva Veda*, a collection of hymns written in India around 1500 BCE, was one of the earliest texts that described rituals based on astronomical knowledge. Texts like this gave the rulers power to know exactly the right day to perform the right ritual to the right god to ensure they achieved their goal, whether that was winning a battle, fathering a son or making sure enough rain fell to water the crops.

## IN TUNE WITH NATURE, AND OURSELVES

For aeons and aeons, humans and animals have lived by the cycles of nature. Our bodies have evolved to be incredibly well tuned into them. How do migrating birds such as swallows know to leave Africa and return to northern Europe in spring? How do they navigate? They do it because they are in tune with and sensitive to the world around them. These days, we humans have found ways of disconnecting from almost every one of these natural cycles. We have electric lighting, central heating and air-conditioning, allowing us to live independently of the outside weather. We have fertilizer, polytunnels and airfreight, meaning the availability of food is no longer bound by the seasons. But what price do we pay for losing touch with the world's natural rhythms?

A keen stargazer is always in touch with the astronomical cycles. Observing the rhythms of the sky helps reconnect us to those of the whole natural world, and being aware of these helps us to notice our own personal rhythms, too. Some people have a lower mood in the winter when the Sun is elusive, while others find their low mood always comes on a Monday morning. Some people like to get up early, others to stay up late. Some find they crave brightly coloured food in the summer and thick, comforting stews in the winter. Some people believe their behaviour or events in their lives are linked to the location of the Sun in particular constellations. The Buddha taught us that all things change. If we can bring attention to our inner rhythms with an attitude of non-judgement and acceptance, we can flow along with the outer rhythms of Earth and the stars more gracefully and sensitively.

# The Moon

I might hazard a guess that every human has stared up at the moon at some point in his or her life – not just acknowledged its presence, but really looked for a good few minutes. It is the most recognizable thing in the night sky. There is something quite magical about seeing a razor-thin crescent on a midnight-blue dusk evening. When it is full, the moon's creamy light can drown out the stars – stars, after all, are pin-like and distant – while in the daytime the Sun is too dazzling to look at; it is only the moon that is large enough and bright enough to be a real presence, hanging there in space.

For most of human history we have not known what the moon is. It appears sometimes as a thin crescent, sometimes a full orb, and at a new place in the sky each night. Sometimes it is not visible at all. All these changes follow a different pattern to the solar cycle, and give us the 30-day lunar month which formed the basis of ancient calendars (in fact, the words 'month', 'measure' and 'menstrual' all derive from the word 'moon'). The moon's cool, silvery light has led people to associate it with the feminine or yin aspects of life: receptive, introverted, indirect and intuitive.

## A SYMBOL OF SELFLESS GIVING

The moon is locked into a synchronous orbit with Earth, meaning it only ever shows us one side. The pattern of darker patches on its surface are called 'maria' (from the Latin 'mare', meaning 'sea', as they were originally thought to.be bodies of water). The pattern of maria look to some people like a face, and to others like animals, spirits or even cheese.

In the Chinese and Japanese tradition, the maria form the shape of a hare or rabbit which reminds us of the Buddha's virtue.

The Buddha taught there are certain threads of consciousness, linked by continuity and the law of karma, which are woven into us when we are born and are passed on when we die, from one rebirth to the next. The ancient Indian text, the *Jataka Tales*, tells the stories of how the different threads of the Buddha's past lives, in both human and animal form, wove their way into him. One such story relates a time when the Buddha was a hare and lived with three friends: a monkey, a jackal and an otter. One day a starving monk came by asking for alms and the three friends very generously offered him what they had caught in the forest that day. The hare had nothing to offer but grass, so gave the ultimate sacrifice: he jumped onto his fire and offered himself as the food. The king of the heavens was so deeply moved by the hare's selfless act that he drew his image on the face of the moon. Because of this, the moon in Buddhist mythology has become a symbol of selfless giving.

It is worth considering for a moment how much you are willing to give to help another in need. Would you be happy to give your time, your expertise, your money or your possessions? Or how about just love? Giving without any regard to what you might get back in return is seen in every culture as a virtue and something we can cultivate with practice.

## WISHING LOVING-KINDNESS

Some years ago, when I was living abroad and apart from my wife, I remember sitting at the window talking to her on the phone late one evening. It was a clear night and the moon was up. When I mentioned to her I was looking at the moon, she went to her window and looked at it too. At that moment I felt connected with her through our shared sight of the moon – the same moon, the same view.

As I write this, there are 7.6 billion people on this planet and roughly half of them are in darkness at any one time. When you look up at the moon, there is a high chance that some of the 3.8 billion people in darkness are also looking up at the moon at that moment. As we share that experience, how would it be to send out wishes of kindness to all those people – whoever they are, wherever they be and whatever they are going through in their life right now? You could try mentally saying, 'I wish for you all to feel safe, to feel at ease, to have good health, to feel contentment and peace.' As you send out these wishes, what is the response in your body? How do you feel? Wishing loving-kindness to friends and family or people who are similar to yourself is relatively easy. But what if that demanding, aggressive client who made you work all weekend was also looking up at the moon in that moment? How would it feel to wish even them good health, contentment and peace?

# Under Mother Nature's Protection

The first time I saw the Northern Lights was on a trip to the far north of Sweden. They certainly danced about, but appeared washed out, faint and colourless. Underwhelmed, I went to bed wondering what all the fuss was about. On the second night, we got a knock on the door alerting us to the lights' reappearance. As we wrapped up for the minus 30-degree cold, our expectations were low – but we were wrong. The lights were amazing. Vivid greens and reds soundlessly swooshed, scintillated, flowed, folded and danced across the sky. We stayed outside until we couldn't stand the cold any more.

It has only been in the last century that people had any clue what produced such beautiful displays. Up until then some cultures viewed them as celestial battles between good and evil dragons breathing fire across the sky. Others saw the lights as spirits of restless ancestors or those who had died violently or tragically in battle.

In fact, what we are seeing are photons of light produced by tiny, electrically-charged particles banging into air molecules as they zoom down through the atmosphere. These particles were emitted by the Sun and on their journey through space have been trapped by the Earth's magnetic field. They are funnelled in towards the Earth's poles by the magnetic force, and as they pass through the atmosphere they interact with atoms of nitrogen, producing green light, and oxygen, giving red light. Think of dust floating in the path of a laser beam; normally we can't see the laser, but when the dust interrupts the beam the path of light becomes visible. Similarly, when we see the aurorae, we are seeing the Earth's magnetic field.

# EARTH'S PROTECTOR

The magnetic field is produced by the Earth's molten iron core, and is incredibly helpful to humans and all life on the planet. Most obviously, we know it through its effect on small strips of iron embedded in plastic cases known as compasses. The iron strip points to magnetic north, which is very useful for navigation. The magnetic field's main effect, however, is to act as a huge shield, protecting us from the harsh weather in space. The Sun gives us light, but also blows off a lot of highly charged and reactive gas particles. Without our magnetic shield, these particles would strip away our atmosphere like a sandblaster (Mars has no atmosphere because its magnetic shield disappeared a long time ago). The ozone layer would be eroded the fastest. We made a start at destroying this layer ourselves a few years ago and now know the terrible consequences. Ozone buffers us from these dangerous particles which, if not stopped, damage (burn) organic matter and can cause cancer. For humans considering interplanetary space travel, this is one of the biggest problems to solve.

Life, in the organic way we know it, can only have evolved because our planet has its magnetic field. Not content with just protecting us from harmful particles, Earth converts them into wonderful light shows. When we develop an awareness and appreciation of this invisible protection, of our world's incredible beauty and interconnectedness, and the delicate balance of life, gratitude seems a natural response. My Zen teacher once remarked that gratitude is the first sign of awakening. That night in Sweden, under the swirling Northern Lights, I felt a huge swelling of gratitude to have the opportunity to witness one of nature's true wonders with my own eyes. I also feel tremendously grateful to the teams of scientists who recognized that the release of chlorofluorocarbons (CFCs) was harming the ozone layer, and to the campaigners and politicians who put the mechanisms in place to ban them.

# GRATITUDE FOR THE WAY THINGS ARE

The practice of gratitude is an important aspect of mindfulness. Right now, I would like to invite you to make a mental note of three things in your life you are grateful for. You can include things as simple as having heating or clean water, or as profound as having a loving partner, caring friends or constantly being under Earth's protective wing. If you do this exercise every day you will notice that some of the things you are grateful for will recur, while others may come and go, but gradually your perspective on the world will change. Gratitude helps us to feel more positive, and replaces our wants, desires and drives with a sense of contentment with how things are in our lives right now. I vow to accept my life for what it is and be grateful for what I have.

# The
# Planets

It is easy to recognize a planet when you see one because it won't be twinkling. Stars twinkle because their single point of light gets distorted by atmospheric turbulence. Planets, however, are close enough to appear as more than just points, and are therefore too big to twinkle. They also slowly move against the background of stars, keeping their own cycle distinct to the Sun and moon. In ancient Greece, astronomers noted the movements of these astronomical bodies relative to the stars and called them planētai, meaning 'wanderers'.

Our solar system had nine planets until 2006, when the powers-that-be updated the definition of a planet to take into account the variety of extrasolar planets (planets outside our solar system) that are being discovered. As a result, Pluto was downgraded to the status of 'dwarf planet'. Pluto is quite small, very far out and impossible to see with the naked eye. At the other end of the scale is Venus. Venus is easily found because it's the brightest planet in the sky. Since its orbit is smaller than Earth's, it always appears near to the Sun (like its fainter cousin Mercury) so can only be seen at dawn or dusk when the Sun is hidden below the horizon. It was the sight of Venus, the morning star, that precipitated the full awakening of the Buddha 2,500 years ago. Some astronomers have proposed that it was a close conjunction of Venus and Jupiter that created the bright beacon that, in the nativity story, led the wise men to the birth of Jesus 2,000 years ago.

Saturn, the next brightest planet, is a popular target for back-garden telescopes. In my experience, people fall into two categories when they first see Saturn in a small telescope: some are thrilled, while others are underwhelmed, having been spoilt by too many amazing images from the Hubble Space Telescope.

# THE LAW OF KARMA

Since time immemorial, the planets have been associated with certain archetypes of the human psyche. Mercury, for example, has come to represent the principle of communication and expression. Venus represents love, harmony and artistic activities, and Mars (easily recognizable because of its bright and slightly reddish colour) represents the principle of energetic force, competition and combat. In the ancient subject of astrology, the configuration of the heavens at the moment of birth was thought to imprint these archetypes on the individual. Astrologers observed the configuration and interpreted how these archetypes would play out in the future.

These days we know that a whole range of factors influence who we are, including the genes we inherit, the kind of parenting we receive, the prevailing attitudes in society and our environment. Some of these factors, like random genetic mutations, are governed purely by the laws of nature and statistics. In the Buddhist and Hindu world view, anything caused by human volition is governed by the law of karma.

In Sanskrit, 'karma' means 'action' and 'vipaka' is the corollary, meaning 'consequence'. According to the law, our present state is influenced by the result of past volitional actions (done by us or others). Generally speaking, it means that if we do beneficial or healthy things, we'll get beneficial, healthy results, and vice versa. However, what makes it complicated is the interaction of volition with the other physical laws of nature and the timescales involved. Some things we or others did a long time ago are still playing out now, and other things (like being hit by lightning) are outside karma and vipaka. It is such a complex intermeshing that the Buddha said that if somebody tried to work it all out, their head would explode (or words to that effect!)

## ACKNOWLEDGEMENT AND HEALING

When we are being mindful, we may encounter a pain, be it a memory, physical discomfort, emotion or perhaps some kind of avoidance. Because it is unpleasant, it is easy to feel like we are doing something wrong. Meditation is about feeling calm and relaxed, isn't it? Not always. Pain arising from the past is a very normal part of the process. When pain does surface, our job is to do our best not to pull back from it, but to acknowledge it, allow it and let it be. Different people have different levels of pain, trauma or suffering to deal with; for some people the ride can be smooth, and for others it can get very bumpy. When we bring a kind awareness to any arising pain and do our best not to push it away, try to change it or even understand why it's there, a deep resolution and healing can happen.

Karma, rather than saying our fate is written in the stars, actually frees us. It says we can take responsibility for our own happiness. It is not about being passive or letting ourselves become a doormat. In any given moment we have a choice to do beneficial and healthy things and set up the conditions for creating our own heaven here on Earth.

# Artificial Stars

Wherever you are in the world, around dusk or dawn you are very likely to see evidence of humanity's increasing use of outer space. If you spot a brightish white light moving at a constant rate across the night sky then it will most likely be a satellite. Occasionally you might see one of these satellites flaring or glinting as one of its flat solar panels catches the light of the Sun at just the right angle, reminding us of their importance.

The year 1957 was the start of the 'space era', when the Russians launched Sputnik 1, followed soon after by Sputnik 2 which carried the first living passenger into orbit (a dog named Laika). Since then, about 6,600 satellites from more than 40 countries have blasted into orbit.

About half of those are still up there today. They perform every function from communication to surveillance (including surveillance of the weather), global positioning, oceanography and astronomy. It is incredible how in just a few decades, our civilization has become so dependent on this orbiting technology.

Our daily weather forecast relies on a very sophisticated network of Earth-imaging satellites looking in the infrared, visible and radio bands of the electromagnetic spectrum. These feed some of the world's most powerful supercomputers, which run simulations of the weather patterns all over the world. Imaging satellites are also the best sources of information on climate change. Environmental protection agencies use Earth imaging to help run and coordinate their campaigns, and relief workers use it for crisis response after major disasters.

Satellites enable phone communication in remote areas of the world and on fast-moving vehicles such as aircraft, ships and trains. Even though your mobile phone doesn't use satellites for its phone or internet connection, it does use them for determining your position via GPS (Global Positioning System) satellites. The advent of GPS has revolutionized the way we drive and find directions to the point where paper maps are now almost redundant. It's amazing how many areas of our lives satellites touch.

## MINDFULNESS FOR THE INFORMATION AGE

It looks like our satellite-fuelled information age isn't going away. In a research paper published in 2017, it was predicted that the flow of data across the globe would grow to 163 trillion gigabytes by 2025, roughly ten times the level of 2015. Moving forward into this kind of world, it is important that we equip ourselves with the tools for dealing with information technology without getting overwhelmed or becoming a slave to it.

Social media, which makes up a good proportion of that global data flow, is designed to catch our attention in a thousand different ways. The result can be a fragmentation of our awareness. Research has shown that between 2000 and 2015 the average Westerner's attention span dropped from 12 seconds to 8 seconds. This is where mindfulness can really help. When we meditate and rest our attention on something, we notice every time this attention drifts off and we bring it back. In this way, mindfulness is actually a practice for training our concentration skills.

Social media can subtly lead us towards narcissism, too. However, with awareness and the right intention, messages on social media can be of great benefit. When posting on social media, it can be helpful to consider what the Buddha had to say about communication. He taught that a statement 'endowed with five particular factors would be blameless and unfaulted'; these were that it 'should be spoken at the right time, in truth, affectionately, beneficially and with a mind of good-will'.

# TECHNOLOGY AS OUR FRIEND

We need to be aware of the possible negative side-effects of our information age. But we can also make use of technology to help us be more mindful. Zen master Thich Nhat Hanh advocates setting an alarm every 15 minutes to remind you to come into the moment. There is now a whole range of guided meditation apps that give access to recordings from many teachers and record how many minutes you have meditated. Other wearable technology allows you to track your breathing patterns through the day, your heart rate and heart rate variability (a measure of your state of calm). All these tools can help us be more mindful if used in the right way.

The next time you are outside on a clear night enjoying the stars and your phone pings, notice the compulsion to check it. Be aware of how your hand automatically moves to your pocket. Notice that impatient yearning to know who is contacting you. Where do you feel it most in the body? How do these feelings change as you stay with them, without looking at your phone? Use that ping as a reminder to reconnect to the moment.

# Polluted Skies

Have you ever been to a place where at night it is pitch dark right down to the horizon, and the entire sky is filled with stars? One of the real joys of the time I spent as an astronomer was stepping out of the observatory into such darkness. But to me there was always an edge of sadness to a view like this. It was only a treat because I lived in a city so light-polluted that on a good night I could only just make out a handful of stars.

Light 'pollution' is the unwanted effects of night-time lighting. All the light that leaks upwards, either directly from badly angled lights or reflected from the ground, illuminates clouds and dust in the air that in turn shine back to the ground. These days many street lights have been fitted with 'full cut-off' shades that direct all the light downwards. When you are on a flight passing low over a city you can easily spot the difference between older street lights that leak upwards (together with many floodlights and security lights) and the newer shrouded lights which are only visible through their dull reflected glow on the road.

The first telescopic observatories were often built on accessible hilltops near to universities or cities. The famous Royal Observatory in England was established in the seventeenth century in Greenwich, just on the edge of the City of London. The site later came to define the prime meridian. As the light pollution grew in the mid-twentieth century, the Royal Observatory was moved to Herstmonceux, near the south coast of England. Then in the 1970s the telescopes were moved to a dark site on a mountain in the Canary Islands off the west coast of Africa.

# THE IMPACT OF THE WAY WE LIVE

As more and more people move into urban areas, the worry is that light pollution will only get worse. It diminishes our quality of life. Children that grow up in cities never get to see the majesty of a truly dark sky. All they see is the dull halo of reflected street lighting and the flashing red-and-white signals of aeroplanes and helicopters. When we truly practise mindfulness, we start to see the impact of the way we live on the world around us. What is your contribution to your area's light pollution? Is there anything you can do to minimize its effect? Do you have any security lights, for example, that could be angled down or shrouded? Light pollution is waste; it costs money and is potentially harmful. The plight of baby turtles in Barbados is just one example. After hatching on the beach on a full moon night, they often get confused on their way to the sea by the lights of nearby towns and end up going the wrong way, wandering onto the road or falling into drains.

In the radio spectrum, the pollution is much worse. TV and mobile phone masts, wi-fi signals and emission from power lines obscure so many of the radio frequencies that governments have had to create radio-quiet zones around some of the world's largest radio telescopes. We have never had such high levels of background microwave and radio emission as we experience today. We don't know the long-term effects to our body's organic tissue of the transmission from our always-connected mobile phone in our pocket.

## LOOKING AFTER OUR WORLD

Practising mindfulness gives us awareness of our actions, our habits and their consequences. What was unconscious, becomes conscious; impulsive reactions turn into chosen responses. We see that every action we take is a choice, and we can choose to live in a way that helps to reduce suffering and harmonizes with our world – or not. Mindfulness also shows us that we are in this together. If someone suffers from the effects of our actions, then ultimately we suffer too. As you read this, do you notice any physical responses or sensations arising as a result? How do these words sit with you? Remember, whatever you find isn't good or bad, it is just what it is. Consider, is there anything you could do differently in your life which would help minimize any suffering caused by the way you live in the wider context?

The practice of mindfulness ultimately connects us to the bigger picture. Our solar system is roughly 4.5 billion years old. As far as we can tell, the Sun has about another 5 billion years to go before it runs out of fuel and fades into forgotten history. How would it be to live in a way that's sustainable, not just for our children or their children, but for humankind's existence on this planet for the next 5 billion years? That would be true sustainability.

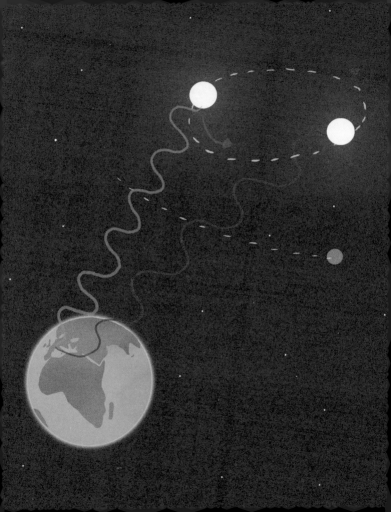

# Through the
# Looking Glass

When you pass light through a prism it splits up into its constituent colours like a rainbow. This is because red light is bent in the prism slightly more than blue light. In the early nineteenth century the German physicist Joseph von Fraunhofer noticed that when light from the Sun shone through a prism it showed dark bands in its rainbow. Much later, people discovered that these dark bands are caused by the absorption of light by certain atoms and chemicals in the Sun's atmosphere. Identifying these bands and measuring how dark they are gave the first indication of what the Sun is made of. William Herschel, a contemporary of Fraunhofer, found that the Sun's light had measurable strength beyond the red end of the colour spectrum and called it infrared light. Physicists later realized that what we see in the rainbow is only a very small portion of an electromagnetic spectrum that extends from radio waves through microwave, infrared, visible, ultraviolet and X-ray, all the way to gamma rays.

Against a bright background we see dark bands in the rainbow, but when we look at light against a dark background we might see certain bright lines. Every atom absorbs or emits light at particular frequencies, just like the skin of every human finger has a different pattern of swirls. Using this knowledge, when we observe light from space split into its spectrum we can identify the make-up of the gas that we're looking at. This technique is called spectroscopy and it is what turns astronomy from simple taxonomy into a true science.

There is more information we can get from a spectrum than just the composition of the source of the light. With a few equations we can work out the temperature, and we can also measure the source's radial velocity via the Doppler effect. Think of the sound of an approaching car. As it gets nearer the sound appears to get higher in pitch, and after it passes it gets lower. In exactly the same way, the colour of light gets bluer when approaching, and redder when receding. This is exactly how Edwin Hubble discovered in the 1930s that the Universe is expanding.

## IMAGES OF OUR UNIVERSE

The stunning images we see of nebulae and galaxies with amazing colours are made using filters that isolate the light from certain regions of the spectrum. For example, if we can isolate the red fingerprints of hydrogen, the green of nitrogen and the blue emission from oxygen, we can combine them into a colour image that both looks beautiful and shows scientifically useful information. Is it actually what we would see if we could get close enough? No, because our eyes perceive things differently to the telescope. Our pupil is only a few millimetres in diameter, whereas the world's biggest optical telescopes are over 10 metres in diameter, and our eyes don't come equipped with narrow-band filters.

## SEEING CLEARLY

So, what do things really look like? What is real? Comparing notes with a fellow traveller on the mindfulness road we find that their experience of things is often very different to ours. Although we don't have filters in our eyes, we do have plenty of filters in our mind. Some people perceive their glass as half-full while others see it as half-empty. Some see obstacles where others see potential; some see astronomy as irrelevant and others see it as the place to study the limits of human knowledge.

The philosophy underlying Zen describes different kinds of 'consciousnesses'. When one of our sense organs (say our eyes) registers an object (for example, the red giant star Betelgeuse) we get the arising of 'sight consciousness'. This is interpreted by our 'mind consciousness', which is coloured by our past experiences, habits and desires. We may either say, 'Wow, it's so red!' or 'That's boring'. Our reality is created when our mind touches what we sense from our environment.

The colouring, or distortion, arises from our unconscious, or what in the philosophy is called the 'seed storehouse'. This is vast and contains all our past experiences, tendencies, cultural conditioning, habits, likes and dislikes.

With the practice of mindfulness our awareness expands so that what was originally unconscious slowly becomes more conscious. Without judgement, we become aware of the seeds in our storehouse – our mind filters – so that we come to know ourselves and experience things more directly. Then, as our lens gets cleaner and we gain the ability to collect more light, we start seeing the Universe in multidimensional technicolor, and realize that what we used to think of as our reality is actually just the tip of the iceberg.

# Time and Space

It is only in the last 100 years that we have started to appreciate just how far away everything is when we look into the night sky. The stars visible to the naked eye are mostly around 10–1,000 light-years away. Even though light travels at a mind-bending 300,000 kilometres per second, it still takes hundreds of years to reach us from some of these 'nearby' stars.

The light from the Sun is already 8 minutes old when it reaches us, and from the outer planets it is a few hours old. The furthest thing we can see with the naked eye is the Andromeda galaxy, whose light left on its journey to us 2.3 million years ago, just as the earliest humans were evolving. I spent my PhD studying a galaxy that is 12 million light-years away, and that's considered the 'nearby Universe'. The furthest object we can find using today's most powerful telescopes is around 13 billion light-years away. Its light has been travelling through space for almost the entire history of the Universe – and just happened to hit the mirror of a telescope and be detected here on Earth.

## THE FOURTH DIMENSION

Space is really a time machine. The further away we look, the further back in time we see. But distance and time are not actually separate; Einstein realized that together they make up the four dimensions of our Universe – what he called 'space-time' (the three dimensions of space, plus the extra dimension of time). One of the questions I often get asked is, 'What's outside the Universe?', or phrased another way, 'If the Universe is expanding, what is it expanding into?' Because the three dimensions of space are inherent to our Universe, the concept of 'outside' is actually flawed. 'Outside' implies space, but space only exists in the Universe. Similarly, another question that is in fact a non-question is, 'What came before the Big Bang?' Time, like space, only exists in the Universe – there is no time without the Universe.

At this point take a moment to tune into your body and mind. Are you still with me? If you are starting to glaze over that's fine. Just let it be, and keep on reading. How do you recognize when you've stopped following the words? What does it feel like? Space-time is a curious thing. Believe it or not, it is the interaction of space-time with objects in the Universe that creates gravity. Gravity is the effect of mass distorting space-time. Imagine space-time as a blanket held out by its four corners. Now put a football on the blanket and see how it sits low in the taut material. Now add a tennis ball and watch how it rolls down towards the heavier football. If the football is the Earth and the tennis ball is us, then we see how gravity gently tethers us to the planet's surface. But remember space and time are two sides of the same coin. Just like the blanket is stretched by the football, so is time. The greater the gravitational field, the slower time appears to pass relative to other parts of the Universe. Einstein called this his Theory of Relativity and it blew people's minds when it was published, and still does. Maybe you feel that way yourself?

## LIVING IN THE HERE AND NOW

When your mind is reeling, it can be a good time to practise mindfulness. At this point our normally overactive thinking mind is preoccupied with cooling its circuits, giving our more direct, experiencing mind a chance to feel and sense. When we are doing walking meditation, we can feel our feet lift up, swing through and come softly to the ground. Knowing about gravity and space-time doesn't help us experience that reality; it hinders it. To directly touch our reality we have to put aside our ideas about it, and realize that these words are just human-made labels to help us understand and relate to our experience. What is time? Hasn't the past already gone and the future not happened yet? There is only now and here. When we remember the past, we do it in the present. Equally when we plan the future we do it in the present. Those light rays that have been travelling through the Universe for 13 billion years know nothing of their history. They arrive now and deliver their information now. All of history, past and future, exists only now.

This is why it is so important how we live right now. It is all we have. Apple co-founder Steve Jobs would apparently ask himself every day, 'If this were my last day, would I do what I'm about to do today?' and if the answer was 'no' for too many days in a row then he would reassess his plans. So, how are you going to live today?

# A Sprinkling of Stardust

As you are reading these words, take a moment to sense your body: its posture, its shape, the space it takes up, the strength of the bony structure, the flexibility and sensitivity of your skin, and your body's sheer vibrancy and aliveness. Being alive in this instant, 13.5 billion years after the beginning of the Universe, is a truly remarkable thing.

When the Universe began with the Big Bang there was nothing but energy. As it cooled, out of this soup came the fundamental particles (quarks, electrons and so on), and later hydrogen (the simplest atom) and a small amount of helium. That was it for 500 million years or so. It wasn't until the first stars formed that any of the other elements came into being.

In the deep furnace of a star, nuclear fusion converts hydrogen into helium to release energy. As a star runs out of its hydrogen fuel, depending on its size, it may fizzle out or become unstable and explode. Fizzling stars in their last gasps for energy begin fusing helium into the heavier lithium, carbon and nitrogen. They are also likely to throw off layers of this newly enriched gas in slow, elegant waves, creating what are known (somewhat confusingly) as 'planetary nebulae'.

Exploding stars, called supernovae, have enough energy to synthesize much heavier elements, from oxygen up to krypton, including all the well-known elements such as sodium, aluminium, silicon and chlorine. Whether stars are fizzling or exploding, the newly made chemicals are blown off into the surrounding space. After some time, they may find themselves in new gas clouds that, in turn, collapse, form new stars, live, die and return more newly made chemicals to space. Our Sun is thought to be a third-generation star that has incorporated all the chemicals made by the stars that came before it.

# WE ARE MADE OF STARDUST

The formation of a star and its planets out of a collapsing gas cloud works like a centrifuge. The light elements (hydrogen and helium) stay in the middle, while the heavier elements spin out to form the planets. The Earth, for example, is composed of 90 or so different chemical elements out of which everything we know is made. There are about $7 \times 10^{27}$ (7 with 27 zeros after it) atoms in an average human body, but 96 per cent of the body is made of just four elements: hydrogen, carbon, nitrogen and oxygen. Most of that is in the form of water – one oxygen atom bonded to two hydrogen atoms ($H_2O$). That means the vast majority of what your body is made of (hydrogen) was created in the moments just after the Big Bang. The rest was forged in the fiery cores of a few stars between about 12 billion and 5 billion years ago. The famous astronomer Carl Sagan was always keen to remind us that everything from the nitrogen in our DNA to the calcium in our teeth and the iron in our blood were once part of the boiling interior of a star. We are literally made of stardust.

Now bring your attention back into your body. Has your posture changed? Notice particularly your teeth. Are they together or apart? Move your tongue around them and sense their shape and surfaces. Your teeth are coated in enamel, primarily a crystalline calcium phosphate. Take a moment to appreciate that the calcium, oxygen and phosphor constituents were formed when lighter elements were fused together in the unimaginable heat of a long-forgotten supernova explosion. It's pretty mind-boggling to think that your teeth were once in a supernova. Now bring your attention to your blood. Can you sense your heartbeat and the flow of blood around your body? Can you remember the last time you cut yourself and bled? Blood contains many elements, including the iron in the haemoglobin that helps bind the oxygen to the cells for transport throughout the body. All of these elements were created inside a star in the distant past.

Modern electronics require much rarer elements. Your phone, for example, uses palladium in the screen, platinum in the storage circuits and gold in the electrical connections. These elements are beyond what even a supernova can make. They are thought to have been created when certain remnants of massive stars called neutron stars collided and exploded. These are some of the most energetic events that can occur in the Universe.

Through everything we see, touch, eat and manipulate, and through the very fabric of our bodies, we are intimately connected with the vast Universe through the billions of years of its evolution. We are the product of generations of stars forming, living, dying and exploding.

# A Photon's Journey

In the deep core of our Sun, hydrogen nuclei fuse to create helium, releasing energy. This energy takes the form of very high-energy photons, or particles of electromagnetic energy, which ping away at the speed of light. Let's jump on board one of these photons and see where it goes.

Outside the Sun's core is a layer called the radiative zone. As our photon enters this layer, the density is so high that the poor photon can't travel more than a few millimetres before it gets absorbed and immediately re-emitted by a gas particle. This happens countless times and the outward progress of our photon is slowed to a veritable crawl. It is nigh on impossible to keep track of where it is and which direction it's going in. The radiative zone is

300,000 kilometres thick and all of this bouncing around means that we'll need to stay with our photon for a few million years before it escapes to the next layer. In this absorption and re-emission process our photon loses much of its energy, sliding from the gamma ray to the visible spectrum. The next layer of the Sun is still 200,000 kilometres thick but it's far less dense, so our proton's journey through it takes only about 10 days. Once it is through, the proton reaches the photosphere, otherwise known as the surface, where it can launch itself out into space.

After having spent millions of years getting through 500,000 kilometres of the Sun's interior, our photon now traverses 150 million kilometres of free space in 8 minutes. At this point it meets the Earth's atmosphere. Dodging space debris and satellites, dust particles and raindrops, our photon zips down towards the ground and strikes the leaf of a lettuce plant. Chlorophyll proteins in the cell of the leaf absorb its energy, allowing the carbon dioxide and water the plant has gathered from its surroundings to be converted into glucose and oxygen. The glucose is used to produce cellulose and other proteins for growth, and the oxygen is released as a waste by-product.

Plants are all transducers of sunlight into chemical energy. When we trace it back, the primary driver of life here on Earth is the energy we receive from the Sun. The next time you eat a salad, consider the lettuce leaf on your plate. Can you see the Sun in that leaf? Can you see the rain and the soil, its seed, the farmer, the delivery driver and the person who washed and prepared it, all in that leaf? Can you see the evolution of life on this planet, the formation of the solar system, the endless stars and galaxies of the Universe in that leaf? We live in a world that is so deeply interconnected. Without any one of these things, the leaf couldn't be here either.

## A FINE BALANCE

If the Earth were a touch closer to the Sun, more photons would hit the surface and it would be hotter; all the planet's water would boil. If the Earth were a little further away, it would be colder, and all the water would freeze. If the carbon dioxide levels in the atmosphere were higher, the greenhouse effect would mean that the Earth's surface temperature would rise out of control. If the Earth were bigger, the surface gravity would be greater and life forms would find it impossible to stand up. The Earth inhabits what has come to be known as the 'Goldilocks zone': everything is 'just right' for carbon-based life to have evolved.

## LET BE, AND LET GO

When we see that life on our planet hangs in this very fine, interdependent balance, it is easy to realize that even the tiniest shift could mean catastrophe. One of the most difficult areas to bring mindfulness to is how we live: our unsustainable need for comfort and convenience and our insatiable desire for something new. Mindfulness shows us that the root causes of this destructive behaviour are our attachment to fame, fortune and power, and our attempts to numb ourselves to our pains and anxieties. The definition of mindfulness is to become aware of what is happening in the present moment without wishing it were different. This simple but often difficult practice is to acknowledge and accept, to let be and let go. If you are carried by your unrecognized wants and aversions, then your actions become selfish and can easily lead to greater suffering. When you can see and accept things as they are, and find peace in yourself, then your actions can arise out of compassion, with a broad, inclusive perspective. You can make decisions and take actions that can benefit all of life on Earth, present and future.

# Is Space Really Empty?

Space is incredibly empty. If the Earth were the size of a tennis ball held in your hand, the Sun would be just over 7 metres in diameter and almost 800 metres away. Neptune, the furthest planet in our solar system, would be a staggering 24 kilometres away. Now, if we made the Sun the size of the tennis ball, the nearest star would be 2,000 kilometres away. That's the distance from one side of Europe to the other, with nothing in between.

In truth, the average patch of interstellar space contains a handful of hydrogen atoms per cubic centimetre. By comparison, on Earth the best laboratory vacuums have around 10 billion molecules per cubic centimetre; even the densest interstellar gas clouds are around 10,000 times less dense than that. To all intents and purposes, space is empty. However, because the distances are so vast, the amount of stuff in space adds up to a lot. The famous Orion Nebula is around 24 light-years across and its total mass is about 2,000 times that of our Sun.

Our galaxy, the Milky Way, is a swirling disc of 300 billion stars and countless gas clouds, with a sprinkling of dark matter and a diameter of 100,000 light-years. Beyond that, the next nearest major galaxy is the Andromeda galaxy, M31, at a distance of 2.3 million light-years. Between us is intergalactic space, in which you might find the odd hydrogen atom here and there if you're lucky. Overall, 99.9 per cent of the Universe consists of intergalactic space, so yes, space is very empty.

## A STATE OF FLUX

When the Buddha examined the Universe from the inner point of view, he also found it to be empty – empty of any inherent, solid, unchanging reality. He said that everything that appears at first glance to be permanent or fixed is actually changing. There are no 'things', only processes. Let us look for ourselves. Find some 'thing' in your vicinity. How long has it been there? What is it made of? What if you go back a hundred or a thousand years? Would it have been in the same form? What about a thousand years in the future? However slow the change may be, the thing is still changing – maybe drying, tarnishing, decomposing or oxidizing. You yourself are also changing. Your physical body, your ideas, thoughts, beliefs, relationships and experiences are all in a state of continuous flux.

By nature, we are fearful of change and uncertainty because we can't control it. That makes us want to believe things are more fixed than they are. But without change there could be no life. If an orchestra played one note for an hour, what would that be like? The whole point of a symphony is that every instrument plays different notes and tones, weaving in and out among each other. That's what makes it beautiful. Emptiness isn't nothingness. It just means nothing stands alone as an independent entity; everything is in a dynamic relationship. Emptiness is potential.

## MINDFULNESS OF CHANGE

When we understand and live from a place where we know that all things are in continuous change, then we can really enjoy life. Have you ever been in one of those restaurants where the plates are huge but the servings are tiny? After a fairly under-filling meal you order an amazing-sounding chocolate dessert. When it comes, it is predictably tiny. You take the first bite and there's a sublime explosion in your mouth. Then you look down and realize you've had half of the dessert already. In your worry that it's going to be over all too soon you barely taste the second bite. You miss the exquisite flavour because you're anxious about keeping hold of the experience. How sad. Mindfulness of change allows you to stay with each bite with 100 per cent presence so that you can enjoy them to the full. You know that these few precious moments of enjoyment are there when they're there, and not when they're not. In the same way, mindfulness of change allows us to weather the difficult, sad or painful aspects of life, again knowing that they will always change. Maybe they will get worse, maybe better, but they will always change. This realization allows us to be with others without trying to fix them into an idea of what we think they should be like, or what we want them to be.

The Universe doesn't try to resist change. That single atom of hydrogen floating in the intergalactic medium never wishes it were part of a dense gas cloud or part of a living being. It knows that in the course of time it may be, or it may not. It just is as it is, in that moment.

# Shooting Stars

One of the most magical stargazing experiences we can have is seeing a shooting star. They are always unexpected and totally unpredictable. No wonder they give us that little feeling of thrill and wonder. They are so fleeting, though, that by the time you see one out of the corner of your eye, it's gone. If you are trying to spot one you need to be looking up with a wide-open attention, and with your awareness on your peripheral vision. This is actually quite calming, because when we feel safe, our vision (and in fact all our senses) softens; when we feel anxious or fearful and become hyper-vigilant our vision narrows. In meditation we intentionally soften the gaze to encourage the body to come into a relaxed and receptive state.

A shooting star, or meteor, is created by a meteoroid (a small rocky fragment of a comet or asteroid) searing through the Earth's atmosphere and burning up. Meteoroids bombard the Earth all the time, mostly in the form of dust or tiny sand grains. At certain times of the year you can see a whole shower of them as the Earth passes through the dusty leftovers of the tail of a long-forgotten comet. If the meteor is big enough to survive the fiery journey downwards then it will hit the ground (or sea) and we call it a meteorite.

## THE THREAT FROM ABOVE

Asteroids are generally much larger (from a metre upwards) and are in orbit around the Sun. Most asteroids are small and harmless, but there are thought to be about 10,000 out there in the solar system that are bigger than 10 kilometres. These are the ones that could do some serious damage to our planet if they hit. Luckily most of them have orbits that don't cross the Earth's path, so the chances of a strike are very slim.

It is thought that the dinosaurs died out so suddenly 65 million years ago because the Earth was hit by an asteroid somewhere between 5 and 15 kilometres in diameter. The impact created an unbelievable explosion equivalent to 10 billion Hiroshima A-bombs, and left a crater 200 kilometres across in Mexico. What did the real damage, though, was not the impact itself but the dust which rose into the atmosphere and darkened the sky around the entire globe for years. We live in the shadow of a catastrophe like this happening again one day, as you'll know if you've ever watched an asteroid disaster film. The chances of being killed by an asteroid strike is higher than you might think: it is estimated at 1 in 700,000, which is much more likely than being killed by a shark or falling coconut, but far less likely than dying in an earthquake, flood or car crash.

## LEARNING TO LET GO

The probability of dying by any means, however, is 100 per cent. It might not be in a disaster, but it'll happen. Most of us live in so much fear of this inevitability, but actually being aware of and acknowledging our mortality is one of the keys to truly living life. If we resist getting old, ill and infirm and live with an underlying wish that we were still young, then we exist in a fantasy world. Living under this fear means we are unable to see the beauty and tenderness of each moment just as it is.

Death is a letting-go process – the ultimate letting go. Maybe it is similar to going to sleep. If we resist sleep (for example, when we're worried about something, or if we're in an unsafe environment) it doesn't happen. We fall asleep every night mostly without problems, and experience shows us that, so far, we have always woken up again. The only difference with death is that we don't have that experience of waking up to reassure us. What happens after death is unknown, and by nature we're fearful of the unknown.

Mindfulness teaches us that all-important skill of how to let go. We learn to allow our thoughts and feelings to arise and pass without getting caught up in them. We learn to let go of wanting things to be different to how they are. We learn to let go of our judgements, of our desires and aversions, our fallibilities, and our loves and hates. If we can let go to that extent, then we're ready to let go of life itself. When we stop holding so tightly onto life then we're truly ready to live. When we're no longer motivated by our own self-centred fears and wants, our perspective can broaden to see what others really need. We can become truly useful, compassionate and caring people in this world. And the world really needs people like that.

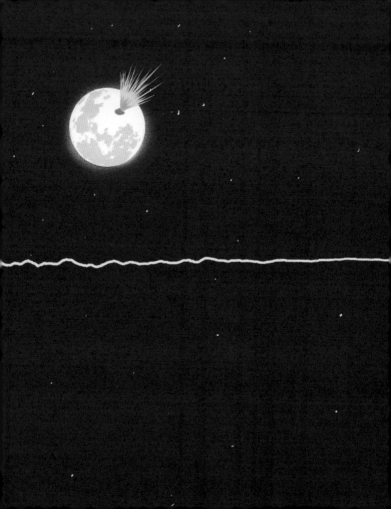

# Entering
# Silence

---

For those of us living in urban areas, it is hard to find quiet. There is the constant background rumble of traffic, sirens in the distance and the whine of aeroplanes overhead. For some, the idea of being in silence is so uncomfortable that they make sure it never happens by always having the television or radio on or music playing in the background. For others that live in the countryside, the silence at night-time can be so complete that you can hear a pin drop.

Unlike our eyes, which we can close against distractions, we cannot close our ears. Mindfulness is therefore often practised in silence. Of course, it is almost never totally silent, but in trying to stay as quiet as possible without any unnecessary talking we minimize any distractions for ourselves and others and help create an atmosphere of space and calm. Staying quiet gives us a chance to feel, to listen and observe the incessant internal chatter – the constant commentary, judgements and reflections on what is going on. However, as much as we might want to, it is not possible to forcibly quieten or clear the mind. The brain secretes thoughts like the skin secretes sweat. Mindfulness is not about how busy or calm the mind is, it is about how we relate to our thoughts. By learning how to accept our mind-state just as it is, things will eventually quieten down, just like a jug of swirling water will eventually settle if you give it time.

Being in silence allows us the space to notice things we might otherwise miss – the wind in the trees, the sparkle of sunlight through the water glass, the tick of the clock or the twinkle of the stars. Looking up at the night sky is a quiet activity. The world is stiller at night-time, and the sky itself makes its slow wheeling procession without a single sound.

# THE QUIET OF SPACE

Sound is made of longitudinal compression waves in a medium like air or water, but space is a vacuum. Space is therefore inherently silent. When you watch a sci-fi film set in space there is almost always a whoosh or hum as a spaceship passes by, and when something blows up you hear the blast of the explosion. All of this, of course, is just artistic license. The tag line for the film *Alien* was true: 'In space no one can hear you scream.' If you were to float in space, you would be struck by just how silent it is. You could be close enough to the Sun to burn to a crisp and still you wouldn't hear its surface boiling and bubbling. A distant supernova could explode and no matter how long you waited there would never be a boom. A comet could crash into the surface of the moon and you would only notice if you actually saw it happen.

It is not strictly true that there's no sound at all in space. In some regions of dense gas, sound can propagate (although it could never reach us here on Earth). NGC 1275 is an active galaxy in the Perseus cluster, 250 million light-years from Earth. It turns out that the supermassive black hole at the centre of this galaxy is slowly vibrating and emitting a sound with a frequency (the time it takes for a single sound wave to pass by) of 9.6 million years. (By contrast, the lowest sounds a person can hear have a frequency of 1/20th of a second.) The sound has caused visible ripples in the dense gas disk surrounding NGC 1275 and is the deepest note ever detected (a B-flat, 57 octaves below the middle keys of a piano).

## FINDING THE HIDDEN SILENCE

Right now, as you read this, take a moment to tune into your sense of hearing. Just listen for a few moments. Try not to label or judge the sounds you hear; just open your awareness to the soundscape around you. Are there any subtler sounds that perhaps take a few minutes to become obvious? Sounds are like thoughts: some are fleeting, while some arise and pass more slowly; some are loud, some quiet; some are pleasurable and compelling, some ugly and aversive.

Now see if you can find the silence underlying the sounds. Take your time. It is always there, like the blue sky is always there behind the clouds. It is there on a busy station platform at rush hour behind all the hubbub and clatter. It is still there when your mind is racing at a million miles an hour with all those to-do lists and ideas. A very traditional image is that of the storm on the ocean. It doesn't matter how wild the wind and rain is on the surface, it is always calm and quiet in the water deep below.

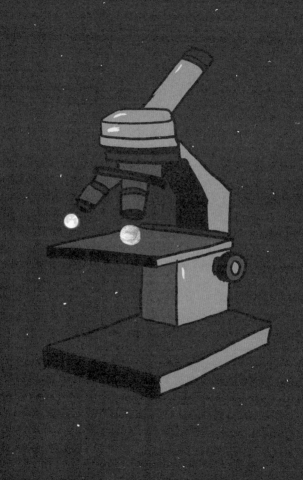

# Zoom Through the Universe

Humans aren't very good at relating to things much smaller or much bigger than our own body size. You might be able to visualize a grain of salt, but go any smaller and our brains go into a daze. We are going to go on a little journey into the very small and the very big. As you read the words that follow, try to notice the moment when you stop relating to what is being described. Notice how it feels. Is it frustrating or exciting? Do you try harder to understand or start skipping over the words?

## JOURNEY TO THE VERY SMALL

A doorway is normally about 90 centimetres wide, give or take. Let's round that to 1 metre and start shrinking our awareness by orders of magnitude (factors of ten). One order of magnitude smaller than you is about the size of your coffee cup. Your phone's SIM card (a centimetre or so in size) is ten times smaller than that. Ten times smaller than the SIM card, at the scale of one millimetre, we find ants and sesame seeds. A tenth of a millimetre is about the size of a human egg cell, and a hundredth of a millimetre is the scale of red blood cells. A thousandth of a millimetre is called a micron. The depth of a groove in a CD is a tenth of a micron. A hundred times smaller than that is the size of a DNA molecule. Going ten times smaller again we get to the scale of smaller, simpler molecules. A single atom is ten times smaller again. If you can imagine going a thousand times smaller still, then we get to the size of an atom's nucleus, itself composed of protons and neutrons. We're now a factor of a thousand billion times smaller than we started. At this scale we can no longer talk in terms of objects. Solidity and location become increasingly meaningless. At this scale there are really only vibrations and energy.

## TO THE END OF THE UNIVERSE

Now let's get back to normal size, and go bigger instead. An order of magnitude larger than us are things like houses and tall trees; ten times bigger again are skyscrapers. Up another order of magnitude is Angel Falls in Venezuela, which is around 1,000 metres from top to bottom. Mount Everest is just under 10,000 metres high, and the final Tour de France stage to Paris is typically a little over 100,000 metres (100 kilometres) long. We now move out to millions of metres: this is, for example, the distance from one side of the United States to the other. The circumference of Earth is a factor of ten bigger, while Jupiter and Saturn are ten times bigger still. Stepping out another order of magnitude to billions of metres we get the size of the Sun, while 100 billion metres is the distance between it and the Earth (known as the Astronomical Unit). The Pluto–Sun distance is a factor of ten times bigger than this, but the solar system extends out a thousand times further to a few million billion metres.

A light-year is around 10 million billion metres, and our nearest star, Proxima Centauri, is about four light-years away. The Orion Nebula is about 20 light-years across, and globular clusters are around 100 light-years across. Dwarf galaxies such as the Magellanic Clouds are another order of magnitude bigger, and galaxies like our Milky Way and Andromeda are a thousand times bigger still. Our local group of galaxies is about a million light-years in size and belongs to a supercluster of galaxies ten times bigger than that. At this scale, galaxies form filamentary patterns with huge empty holes between them, with typical sizes of 100 million light-years. Going a factor or ten bigger gets us to the size of the entire observable Universe and there is no way of knowing what's beyond that.

## QUIETING THE THINKING MIND

Now you have zoomed in and out through the Universe, consider at what point did you stop relating to the descriptions? What we relate to and perceive as a human is so incredibly limited compared to the vastness of the whole Universe. Astronomy is the best teacher of this – our mind boggles at almost every turn. Some (maybe most) of the aspects of our Universe that we discover are totally ungraspable, no matter how hard we try.

The 'boggle-point', as it were, is actually a very interesting moment. As your intellectual, thinking mind can no longer grasp what it is being given, it essentially gives up. This isn't a bad thing in any way, as it gives our experiencing, sensing mind an opportunity to see reality as it is, unmediated by thoughts or concepts. So, the next time you're looking up at the night sky considering the vastness of it all (remember, it is around 40 million billion metres just to the nearest star), let your mind boggle and immerse yourself in the visceral, felt experience of being part of this wondrous Universe.

# The Most
# Exquisite Jewels

After I had been an astronomer for some years, I found I became numb to the exquisite beauty of the photographs of space captured by the latest telescopes. In the midst of this detachment, however, I remember one image that struck me like a thunderbolt from heaven. It was of the young, star-forming nebula NGC 346 in the Small Magellanic Cloud (one of our closest neighbouring galaxies), captured by the Hubble Space Telescope.

The image looked to me like a myriad of beautiful jewels nestling in a blue, glowing silk pillow. Most of these jewels are actually young, massive stars which are burning bright and hot. The radiation and particle emission from these powerful stars are gradually blasting away the remaining gas in the cloud, creating the wonderful, wispy backdrop. The blue colour (which was probably shamelessly dialled up in the rendering of the image) represents hydrogen and oxygen emission from the gas as it is lit up from the inside by the stars.

## AN INTERCONNECTED WEB

The majesty of NGC 346 is a worthy example of the incredible vision of the cosmos laid out in the Buddhist *Avatamsaka Sutra*. This 'Buddha's eye view', which describes the depth and detail of the Universe as seen by a Buddha, was quite an undertaking, so it is little wonder that the sutra is one of the longest spiritual texts ever written. NGC 346 reminds me in particular of the sutra's metaphor of Indra's Net. Indra's Net is a three-dimensional net which permeates the Universe, with a multi-faceted jewel positioned at each vertex. Each face of each jewel reflects the light from every other jewel in the net, creating a vast interconnected, interrelated web. This metaphor helps us see how every grain of sand on the beach or every atom in our own body contains the essence of the entire Universe. Zen master Hakuin wrote of 'Ten million Universes in a dewdrop on a hair-tip; The entire Cosmos in a fleck of foam on the sea'.

The atoms making up your body are dependent on the Universe being exactly as it is and having arisen as it has. Your body, this book and this writing are dependent on every single event and process that has formed the Universe, right back to the beginning of time. Zen master Thich Nhat Hanh likes to call this 'interbeing'. You and the page in front of you 'inter-are'. The paper didn't just come from a paper mill. The metaphor of Indra's Net helps us to see that it arose because of everything that has ever happened in the entire history of the Universe. The words you are reading contain the Big Bang, the evolution of the stars and galaxies, the formation of our solar system, all animals that have ever lived, the care of countless parents for their children, and even my lunch today. We all inter-are.

Nothing is a fixed object existing independently, because everything is 'interdependent' on everything else. In reality, cause and effect are not actually separate. A cause must, at the same time, be an effect, and every effect must also be a cause of something else. Eating the grass causes the cow to stay alive, but the cow's dung nourishes the grass and is part of the cause of the grass growing. Nothing exists in and of itself. You yourself don't exist as a separate thing, and that includes your ideas and thoughts, perceptions and identity. When we see that 'I' am made from a whole Universe of 'non-I' elements then we might start to understand that our normal perception of self, of 'me' and of 'mine' is really just an illusion.

## THE STARS CONTAIN US ALL

That image of NGC 346 can now take on an even more beautiful aspect. All those star-jewels actually reflect and contain every other star-jewel and galaxy-gem in the entire Universe. They contain you, me, love, warm summer days, chocolate ice cream, a clogged drain, crumbling monuments and everything that has ever been. We are both an effect and a cause of that exquisite glowing pillow of glinting sapphires and rubies 10,000 light-years away. When we look up into the night sky it may seem like everything is remote, distant and far removed from our daily life. But it is the contrary. We and all the stars inter-are. We could not exist without them, and they could not exist without us. That is what makes stargazing so amazing.

# How Many Stars Are There in the Sky?

---

In Zen, we often meditate on a question. The question people often start with is 'who am I?' Zen has about 500 such questions, or 'koans', but they all point at the same answer – the answer to 'who am I?' Some koans are about finding this wisdom, some are about how to apply or live that wisdom, and some are about expressing that wisdom. Some are straightforward, while others are a little more obscure.

The koan that people have traditionally started with down the centuries is Zen master Joshu's 'mu'. Joshu was a Chinese teacher from the Tang Dynasty around 800 CE.

One day, a monk came to him and asked if a dog had Buddha-nature (that is, the nature of complete OK-ness just as things are). The Buddha taught that all beings have Buddha-nature, so the monk was probably expecting a 'yes' answer to his question. Joshu instead answered with the word 'mu', which roughly translates as 'not'. Joshu had seen that the monk had been caught in his conceptual question about a dog's true nature. By answering so unexpectedly, Joshu shook the monk out of his concepts and ideas and into direct experience. In meditation, we look into this 'mu'. Where was master Joshu coming from when he answered in this way? What truth was he pointing at?

## JUST START AND KEEP GOING

I have heard it said that the koan system of Zen is one of the hidden wonders of the spiritual world. It is a complete education system in enlightenment, designed to take you through the gateless gates and hidden barriers of the mind. One of the koans that is about the practical application of wisdom asks, 'How many stars are there in the sky?' Try this koan the next time you are out on a clear night. Stand or sit comfortably with an upright posture; soften your face and notice your breathing. From your view of the sky, let this question fall through your body into your belly. How many stars are there in the sky? Now let anything that comes in response to that question arise. If you get caught up thinking about concepts, ideas or theories, notice that and come back to the question. How many stars are there in the sky?

When we are faced with a seemingly insurmountable task, what do we do? Standing at the bottom of the huge cliff face, we can easily feel paralyzed by the sheer enormity of it. Where do we start? How do we start? Therein lies the direction of this koan. The only way is to just start and keep going. Don't look up at the whole cliff face in an anxious daze; look for the first foot- and hand-holds, and start climbing. As the saying goes, 'A journey of a thousand miles starts with the first step.' Imagine you have been asked to write a textbook about a subject you know well.

Where do you even begin? The answer, again, is to just start and keep going. Begin with the chapter headings, then add the section headings, then the subheadings and a few broad bullet points. Gradually more bullets can be added and then expanded. Slowly the notes turn into prose and before you know it you'll have finished a chapter.

## BEGIN WHERE YOU CAN

When we look out at the world, the mass of suffering we see with our eyes, in the news and on social media also seems like an insurmountable cliff face. There is poverty, illness, conflict, corruption, war, discrimination and prejudice everywhere. In the face of all this, how can we possibly help? How can we hope to make a difference? The only realistic thing we can do is start where we can. Of course, we can't tackle child poverty single-handed, but we can put a donation in the local food bank. There is no way we can deal with the widespread plastic pollution of our oceans, but we can choose to take a reusable bottle with us, rather than buy bottled water. If we don't have the time or resources to help out at a soup kitchen at a local homeless shelter, we can do something smaller and more manageable; we could simply decide to say hello to the homeless person we walk past rather than ignore them.

Faced with an overwhelming situation it is important we don't end up in denial, avoidance or paralysis. So, the next time you find yourself in one of these unhelpful states, remember to start where you are and do what you can, however small or insignificant that may seem. Remember, a few flaps of a butterfly's wings can change the course of a tornado on the other side of the planet. So how many stars are there in the sky? Just start counting them.

# Is There Anybody Out There?

---

What is the meaning of life? Of course, there is the scientific definition of what life is, but that doesn't help us with the meaning. For that, we study the self – because we are life. As Zen master Dogen said in the thirteenth century, 'To study the Buddha's teachings is to study the self. To study the self is to forget the self.' When we forget the self we end up in a place where there is no longer any sense of separateness between anything. We gain a felt, experienced, 'knowing' of what life truly is. We become life.

We are life and Earth is teeming with it, but what about 'out there'? This is one of the primary motivations for continuing to explore space. Do you think there is life on other planets? I always found it reassuring that the Buddhist teachings explicitly talk about 'worlds beyond as many worlds as there are grains in sixty-two Ganges Rivers... each with their own Buddha and hosts of attendants.'

## THE GREAT UNKNOWN

To calculate the odds of finding intelligent alien life, astronomer Frank Drake came up with an equation in 1961. He said that it depended on seven things: (1) the rate of formation of stars suitable for the development of life (some stars are very unstable or violent), (2) the fraction of those stars with planetary systems, (3) the number of planets per star with an environment suitable for life, (4) the fraction of those suitable planets on which life actually appears, (5) the fraction of those on which intelligent life evolves, (6) the fraction of civilizations that develop technology which we could detect, and (7) the length of time those civilizations exist for. Unfortunately, at the moment we have almost no idea how to estimate anything after the first two variables.

What is helping us get a handle on the second and third points is the discovery of planets outside our solar system, known as extrasolar planets, or exoplanets. This is one of the most exciting developments in astronomy in the last 30 years. It all started in 1995 when regular dips in the brightness of the star 51 Pegasi b were observed. Since then we have discovered over 3,500 exoplanets, mostly thanks to the Kepler Space Telescope that was launched in 2009.

So far it looks like the number of stars in the galaxy with planetary systems is quite high (point 2). A planet that's suitable for life (point 3) is defined as having liquid water on its surface (orbiting in the so-called 'Goldilocks zone'). Half the number of exoplanets so far discovered inhabit the Goldilocks zone. But we have no way of knowing the likelihood of life forming even if there is liquid water. On Earth, life started pretty much as soon as it could, but you can't measure a trend from one data point. Making educated guesses for the unknowns, our current estimate for Drake's equation gives us about one planet with intelligent life per galaxy. So, in our galaxy, we're it!

# COMPASSION FOR ALL LIVING THINGS

Life is incredibly rare and incredibly precious. We therefore have a responsibility to do our best to foster the flourishing of this life wherever we can. One of the central moral tenets of all the world's religions is the commitment not to kill. In Buddhism, finding out what that actually means is seen as part of one's ongoing practice. Of course, every time we step there's a possibility we will kill an ant or a worm, and driving in the summer we certainly kill many insects with the front bumper. Killing is a part of living. We all need to work out where we draw our line.

Jain monks are strict vegans and sweep the ground ahead of them to make sure they don't step on anything. Some Buddhist monks eat meat (including the Buddha himself), and in Christianity 'do not kill' is largely confined to mean other humans. By developing an awareness of the suffering caused by killing, we see how dependent everything is on everything else. We see that it benefits all when we act with compassion and out of a wish to protect the lives of people, animals, plants and all of nature. When we use a disposable fork or buy a mobile phone we might think it has got nothing to do with killing, but that may not be true. A plastic fork may end up discarded in the ocean where it will be eaten by a fish that chokes on it. The rare metals needed for a phone may be mined by a company sponsoring war. Living without killing is very difficult. Intention and mindfulness are key, but it is important to regularly reassess what living without killing means to us and not to get too hung up on it. As Zen master Thich Nhat Hanh says, trying to be non-violent is like looking at the North Star in order to go north. We do not intend to arrive at the North Star itself, but we use it to point us in the right direction.

# The Overview Effect

As an astronomer, people often asked me if I would like to go into space. Sadly, astronomers and mindfulness teachers are pretty much the last people to be invited to become astronauts. Astronomers might help design the missions and orbital trajectories while mindfulness teachers might help the crew focus and deal with stress, but in space you really need pilots and engineers. But if someone were to offer me a free trip into orbit, I certainly wouldn't say no.

Past spacefarers have repeatedly reported feeling a deep sense of cosmic connection when they look down on the Earth from afar. This life-changing new perspective has been dubbed 'the overview effect'. In 1969, Apollo 9 astronaut Rusty Schweikart was in orbit, going around the

Earth once every hour and a half, when he realized how deeply bound up his identity was with this tiny ball of rock below him. Apollo 14 astronaut Edgar Mitchell – part of a crew which landed on the moon in 1971 – described how, looking back at the Earth from space, he felt a profound sense of connectedness. He realized that each and every atom in the Universe was connected, and understood that all the humans, animals and systems on the planet were a part of the same whole. 'You develop an instant global consciousness,' he explained, 'a people orientation, an intense dissatisfaction with the state of the world, and a compulsion to do something about it.'

More recently, International Space Station astronaut Ron Garan described our planet from orbit. He saw it as 'indescribably beautiful', a living, breathing and extremely fragile organism, and realized how the 'little paper-thin layer [over the surface] is all that protects every living thing on Earth'. When you imagine seeing huge swathes of Earth in your field-of-view, or perhaps even the whole planet, you can understand why Garan used the phrase 'paper-thin'. If the Earth were the size of a football or basketball, the atmosphere would be about the thickness of a thin sheet of plastic wrapped around the ball.

## AT ONE WITH THE UNIVERSE

Astronomer Carl Sagan famously reminded us that underneath this delicate wrapping lives everyone you love, everyone you know, everyone you ever heard of and every human being who ever was. Anything that has ever mattered to us exists here on 'a mote of dust suspended in a sunbeam'. If you have access to the internet nearby, look up a photo of the Earth. Take some time to contemplate it: that beautiful blue-green marble suspended in space. See the continents and the cloud patterns, the enormous oceans and the polar ice caps. That's home.

Viewing our tiny, isolated planet like this, it is possible to fall into a nihilistic black hole. If we're so small and insignificant then why should anything actually matter? Surely the Universe doesn't care how polluted the planet gets or whether I live or die? The trap is set when our world view is oriented around 'little me' versus the 'big hostile Universe'. With that mindset, we are always bound to lose. But it is not the true picture. You and I are not actually separate from the Universe; we are the Universe. We are like two waves on the sea with distinct identities and characteristics, but actually formed from the same body of water. Does the right hand care if the left hand cuts itself? Yes, because they are both part of the same whole. Does the Universe care if someone dies? Yes, because we care.

## SPACESHIP EARTH

The nineteenth-century economist Henry George first articulated the idea of planet Earth as a 'well-provisioned ship, on which we sail through space'. Whilst at sea, a ship is a closed system – what is on-board is all there is. The Earth is like that. We only have the resources we have, but nowadays we appreciate that these provisions are not endless. Seeing our home as 'spaceship Earth' can help us realize that we are all in it together. It becomes obvious that we all need, as Sagan says, to 'create a planetary society with the united will to protect this "pale blue dot"'.

The reality is, however, that we live in a deeply unequal world with unsustainable economic and industrial growth. Currently just 1 per cent of the population owns 50 per cent of the world's wealth. You may not be in that 1 per cent, but if you are reading this book the chances are you live in reasonable plenty. I know we never asked to be born into comfort and prosperity, but having resources gives us opportunities. We, unlike those living in abject poverty, can make choices as to how we live. To me, having prosperity means we have a responsibility to use it wisely. How will you use yours?

# Bringing It Back
# to Earth

In an application form for a big astronomy grant there is always a box asking how the research proposal is important for the general public. Of course, the bulk of astronomical research is not directly relevant to everyday life. It doesn't change how we live to know how a spinning protostar causes magnetically confined jets to blast through its gaseous cocoon as its core begins nuclear fusion. But understanding how stars form and how the fusion reaction works has led us to realize that our very bodies are made from elements which were once synthesized in the cores of ancient stars.

Our bodies are the product of astronomical processes that have taken place since the beginning of time. The fine balance of forces and energies put in place by the Big Bang,

13.5 billion years ago, have allowed our planet to evolve and sustain life. If the strong nuclear force had ended up being only five per cent weaker, stellar nuclear fusion could not have happened at all and the Universe would just be one large cloud of hydrogen.

Knowledge like this helps us broaden our perspective and appreciate the intricate complexity of the Universe. It enhances our experience of the present moment. Experientially we see that we are the Universe. I am those ancient exploded stars, just like I am the sandwich I ate for lunch yesterday. The sandwich became me and I am it. A monk once asked Zen master Joshu what the meaning of Zen was and Joshu replied 'the oak tree in the garden'. Simple as that! The oak tree contains the essence of the acorn from which it grew, together with every other oak tree that ever has lived. It contains all other plants and amoebas, planets, stars and galaxies that have ever been. When we look at the tree with wide eyes and an open mind, that's what we see. That's what Zen and yoga and mindfulness are really pointing at.

# THE UNIVERSE IN A TINY ATOM

We start by bringing awareness to the breath and to other sensations in the body as we sit or move. This helps to train the mind to stay focused and receptive. Through bringing an attitude of curiosity, awareness and acceptance to our experiences we learn how to deal with difficulties such as anxiety and pain. We see how wanting things to go 'our way' only leads to suffering and discontent, and the only response to that realization is to let go. When the mind stabilizes and we stop trying to swim against the flow of life, then there is space to start enquiring into the nature of our experience. Is it fast or slow, heavy or light, full of fear or joy, settling or unsettling? Is it pleasant, unpleasant or neutral? Who is this person that's asking? What am I? What is this thing I call life? What is my relationship to the world? Where did I come from? What is the Universe? Introspection meets astrophysics and vice versa.

We sit and contemplate the oak tree in our garden and suddenly we see the whole kaleidoscopic, holographic, interconnected Universe opening up before us. We see how the tiniest atom contains the essence of the biggest galaxy supercluster, how a moment contains an aeon, and how an aeon can flash by in an instant.

## ACCEPTING THE HERE AND NOW

I originally studied space to escape real life. But the truth is we can't escape. We can only put off facing it for a time. One of the founders of secular mindfulness, Jon Kabat-Zinn, titled one of his books *Wherever You Go, There You Are*. How true. If you want to change your life, reading about mindfulness isn't enough. You need to practise it regularly. It is like learning a musical instrument – a little every day and you'll be playing Mozart in no time. Just 10 minutes or so spent following the movement of your breath with focus and curiosity is all you need. When we come to face ourselves and look at reality just as it is, time and again, something begins to shift. We gradually become lighter, more content and more at home wherever we are.

Catching sight of the night sky (clear or otherwise) can be our little reminder to come home – returning from whatever journey we've been on in fantasy land and arriving in the reality of now. We sense our breath gently moving in and out, our feet being held to the surface of the planet, and the vast expanse of the Universe wheeling above us. We are here and now; we're home.

# ACKNOWLEDGEMENTS

My sincere thanks go to Monica Perdoni, my commissioning editor, for inviting me to write this book; to Tom Kitch and his editorial team for helping refine the book; to the Leaping Hare Press design team and Lehel Kovacs for his beautiful artwork.

And a particular thanks goes to my Zen teacher Julian Daizan Skinner for his infinite patience, wisdom and guidance over the years.